Why Science Matters

Predicting the Effects of Climate Change

John Townsend

Heinemann Library
Chicago, Illinois

Customer Service 888-454-2279
Visit our website at www.heinemannraintree.com

Editorial: Andrew Farrow, Megan Cotugno, and Harriet Milles
Design: Steven Mead and Q2A Creative Solutions
Illustrations: Gordon Hurden
Picture research: Ruth Blair
Production: Alison Parsons

Originated by Heinemann Library
Printed and bound in China by Leo Paper Products

ISBN: 978-1-4329-1839-2 (hc)
ISBN: 978-1-4329-1852-1 (pb)

13 12 11 10 09
10 9 8 7 6 5 4 3 2 1

Library of Congress Cataloging-in-Publication Data
Townsend, John, 1955-
Predicting the effects of climate change / John Townsend. -- 1st ed.
p. cm. -- (Why science matters)
Includes bibliographical references and index.
ISBN 978-1-4329-1839-2 (hc) -- ISBN 978-1-4329-1852-1 (pb)
1. Climatic changes--Risk assessment. 2. Climatic changes--Effect of human beings on. I. Title.
QC981.8.C5T69 2008
363.738'742--dc22
2008014309

Acknowledgments
The publisher would like to thank the following for permission to reproduce photographs:
© Alamy/Photodisc p. **24**; © Corbis pp. **25** (Barry Voight/Pennsylvania State University), **34** (Reuters), **38** (Ashley Cooper), **39** (NASA), **41** (Made Nagi/epa); © Creatas p. **29**; © Digital Vision p. **29**; © DK Images p. **45** (Craig Knowles); © Getty Images pp. **22** (NEC), **33** (Taxi); © NASA p. **23**; © naturepl.com p. **17** (Ingo Arndt), **31** (Martha Holmes), **37** (Jurgen Freund); © Pearson Education Ltd p. **44** (Peter Morris); © PhotoDisc pp. **5** (StockTrek), **29**, **29** (Photolink); © Punchstock/Fancy p. **29**; © Rex Features/Action Press p. **35**; © Photoshot pp. **27** (Newscom), **36** (NHPA), **42** (UPPA); © Science Photo Library pp. **8** (NASA GSFC), **9** (Ge Astro Space), **12** (Sheila Terry), **15** (Philippe Psaila), **16** (Karim Agabi/Eurelios), **20** (Alexis Rosenfeld), **26** (David Nunuk), **32** (Mike Boyatt/AgstockUSA), **46** (Victor de Schwanberg); © Science Photo Library/NASA pp. **19** (top and bottom), **30**. Background images supplied by © Istockphoto.

Cover photograph of parched earth reproduced with permission of ©Corbis (Theo Allofs). Background images supplied by ©Istockphoto.

The publishers would like to thank Sally Morgan for her invaluable assistance in the preparation of this book.

Contents

Some words are printed in bold, **like this**. You can find out what they mean in the glossary.

Our Changing Climate

Scientists look for answers to all kinds of questions. Sometimes they gather information in the hope of solving some of the world's problems. A big problem facing all of us today concerns what is happening to our planet. Earth is showing worrying signs that it may change forever. Its climate is changing like never before. **Global warming** is melting ice, raising sea levels, causing floods and droughts, and threatening plant and animal species. Scientists predict that these problems will continue, but for how long, and at what rate? And what might the long-term effects be?

What is climate?

The weather describes the state of the **atmosphere** at a particular place and time. The temperature, amount of sunlight, wind speed, and rainfall are all weather conditions. The climate is the average weather affecting a region over a long period of time. Climate change is not about getting the occasional day of unusual weather. It is about a fundamental shift in long-term weather patterns that could mean massive changes for everyone's way of life.

Why such concern about climate change?

Climate change is nothing new. Different parts of Earth have been heating and cooling over millions of years. In fact, some scientists believe it was the effects of natural climate change that eventually led to the extinction of dinosaurs and many other species. Many **ice ages** have come and gone, with some extreme temperature swings. If the planet survived climate changes in the past, why is there such concern now?

In the past, climate change has happened as a result of natural events. It is only in the last 150 years that humans have been affecting natural processes so greatly by polluting the world. Some people say that an increase in the melting of the world's ice is proof alone of global warming. The most authoritative independent review by climate scientists, the Inter-Governmental Panel on Climate Change (IPCC), has stated clearly that climate change is happening, that it is happening because of human activity, and that drastic and urgent action is needed to tackle it. Now scientists are trying to find solutions to climate change and prevent catastrophic effects for life on Earth. What more proof do we need that science really matters?

Earth's climate has always been changing, but never as quickly as it appears to be changing now. Scientists are warning that humans are causing this climate change. We need to act fast to save our planet.

Earth Science

Scientists have always sought to understand how Earth sustains life. Other planets in the solar system are too hot, too cold, or without the right balance of gases. Earth has just the right conditions because of its distance from the Sun and the angle of its tilt.

Influences on Earth's climate

Earth is powered by the Sun. This **solar energy** drives all of its systems and cycles. One of the key cycles affecting weather and climate is the water cycle. Without the Sun's power, the water cycle would stop (see box below).

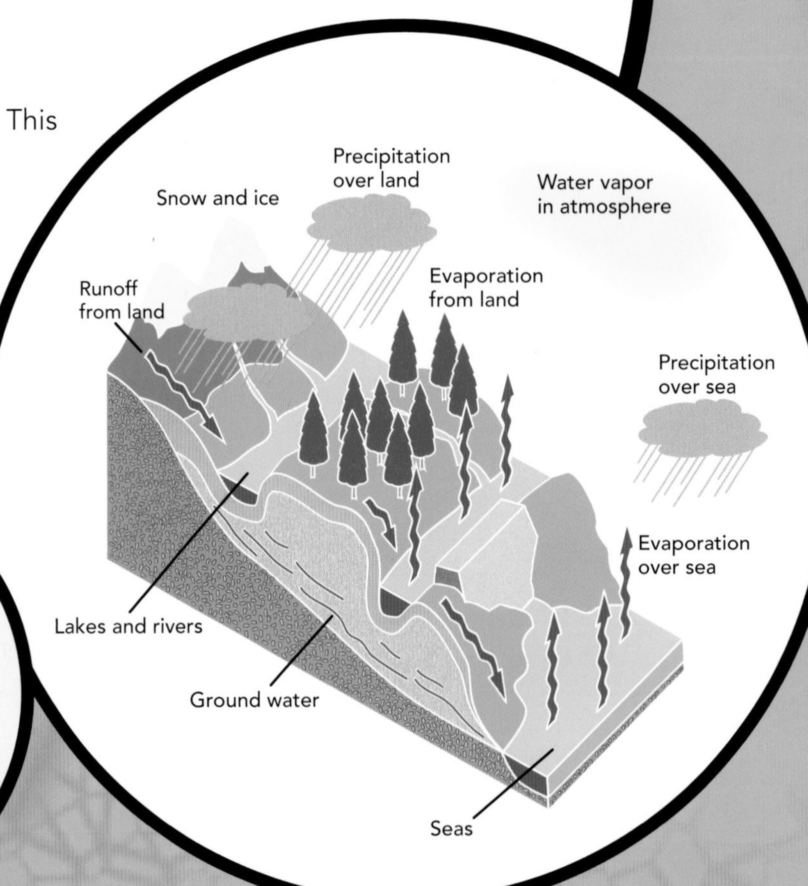

The Sun's power to move water and gases around the planet is one of the main influences on climates, environments, and **ecosystems**.

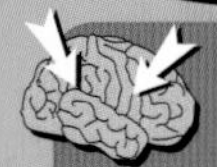

THE SCIENCE YOU LEARN: THE WATER CYCLE

As the Sun beats down, it warms oceans, rivers, and lakes. Warmed water rises into the air as water vapor in a process called **evaporation**. A tiny amount of moisture is also released from trees and plants. This is **transpiration**. As water vapor rises into the atmosphere, it cools and forms liquid water droplets in a process called condensation. The droplets group together into clouds, and when they are too large to remain in the atmosphere, they fall as rain, snow, or hail. These processes are part of the water cycle, which continuously moves water around Earth.

The greenhouse effect

Our planet's climate is far from simple. Many factors influence it, from events on the Sun to the growth of microscopic plants in the oceans. The gases that make up Earth's atmosphere, and in particular the **greenhouse gases**, are vital in affecting Earth's temperature.

Earth's mixture of greenhouse gases acts like a blanket, keeping the planet warm by trapping solar energy. Without those gases, Earth would be on average 33°C (60°F) cooler. The greenhouse gases include carbon dioxide, methane, and nitrous oxide. Other gases, such as water vapor, also act as greenhouse gases. As we shall see on pages 10–11, humans have been increasing the level of carbon dioxide in the atmosphere.

Greenhouses work by trapping heat from the Sun. The glass of a greenhouse lets in sunlight but does not let all the heat escape. This causes the greenhouse to heat up. Greenhouse gases in the atmosphere act just like the glass in a greenhouse. Sunlight enters Earth's atmosphere, passes through the greenhouse gases, and hits Earth's surface. Much of this energy bounces back into the atmosphere. Some escapes up into space, but some is trapped in the atmosphere by the greenhouse gases, causing Earth to heat up.

The greenhouse effect is very important. Without it, Earth would not be warm enough for humans to survive. However, if the greenhouse effect becomes stronger, Earth warms up more than usual. Even a little extra warming will cause major problems for humans, plants, and animals.

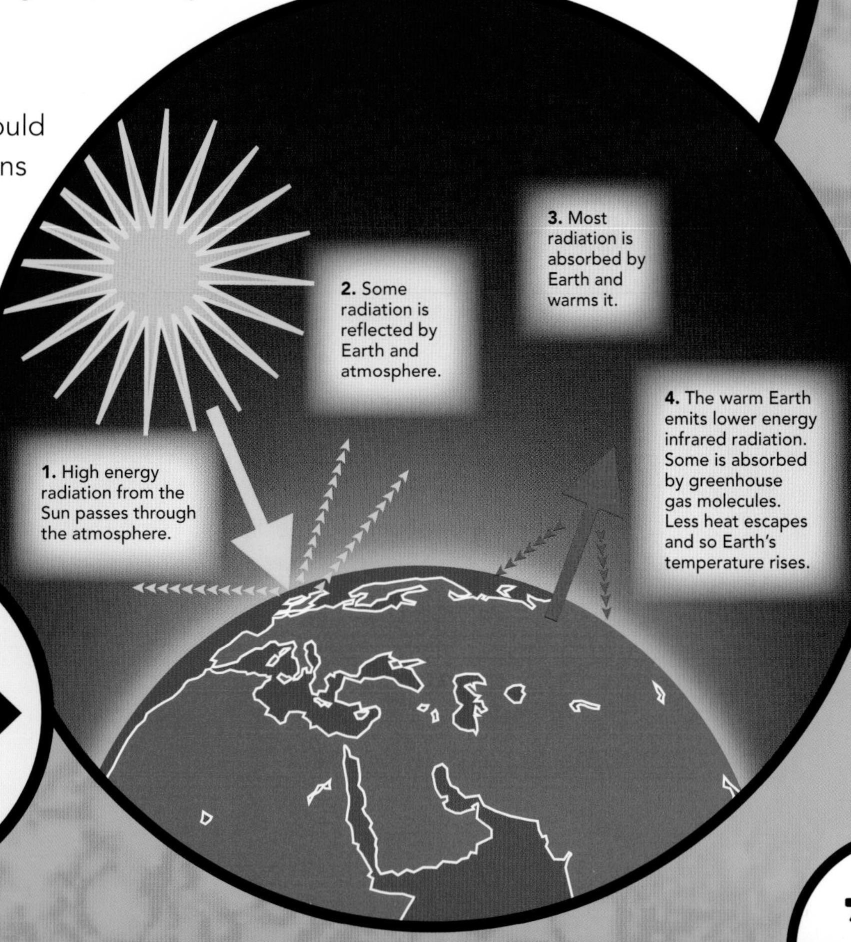

1. High energy radiation from the Sun passes through the atmosphere.

2. Some radiation is reflected by Earth and atmosphere.

3. Most radiation is absorbed by Earth and warms it.

4. The warm Earth emits lower energy infrared radiation. Some is absorbed by greenhouse gas molecules. Less heat escapes and so Earth's temperature rises.

The diagram shows how the greenhouse effect works.

Climate

The Sun and its energy not only affect Earth's temperature, rainfall, and atmosphere, but also its oceans and winds. These factors are all interlinked. Conditions in one part of the world can influence the climate thousands of miles away, particularly where ocean currents and wind patterns are concerned.

The ocean

Oceans cover about 70 percent of Earth's surface and store huge amounts of heat. The top few feet alone store as much solar energy as the entire atmosphere. Scientists have discovered that winds passing over the oceans carry moisture and warmth to nearby land. Winds blowing over warm ocean currents, such as the Gulf Stream in the North Atlantic Ocean, can raise temperatures in normally colder places farther away.

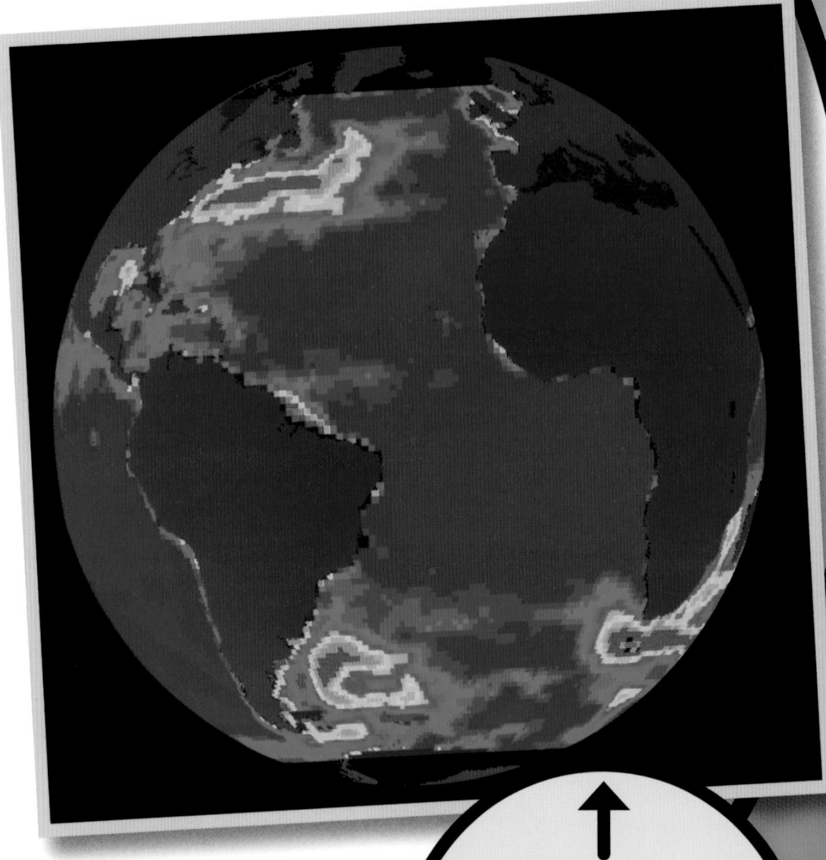

Ocean currents in the Atlantic Ocean. The red patches represent the main warm-water currents that influence climate.

For decades, scientists have studied and traced ocean currents by throwing labeled bottles into the water and plotting their routes—often halfway around the world. Bottles from the Arctic have ended up in all sorts of places, including Brazil. This proves that conditions in the Arctic can affect any part of the planet. In addition, solar energy is reflected by sea ice and ice sheets in the Arctic and Antarctic, helping to regulate Earth's climate.

The sky

Solar energy, the rotation of Earth, variations in Earth's surface, and the oceans control our weather systems. However, it is the planet's upper atmosphere that receives a lot of attention from scientists. A problem in the upper atmosphere could permanently disturb weather systems. As well as providing the greenhouse effect, the layers of gas in the atmosphere perform another function. The **ozone layer** shields the planet's surface from the Sun's harmful **ultraviolet radiation**, and makes life on Earth possible. Human activity has damaged the ozone layer (see panel below), allowing more ultraviolet radiation to reach Earth's surface.

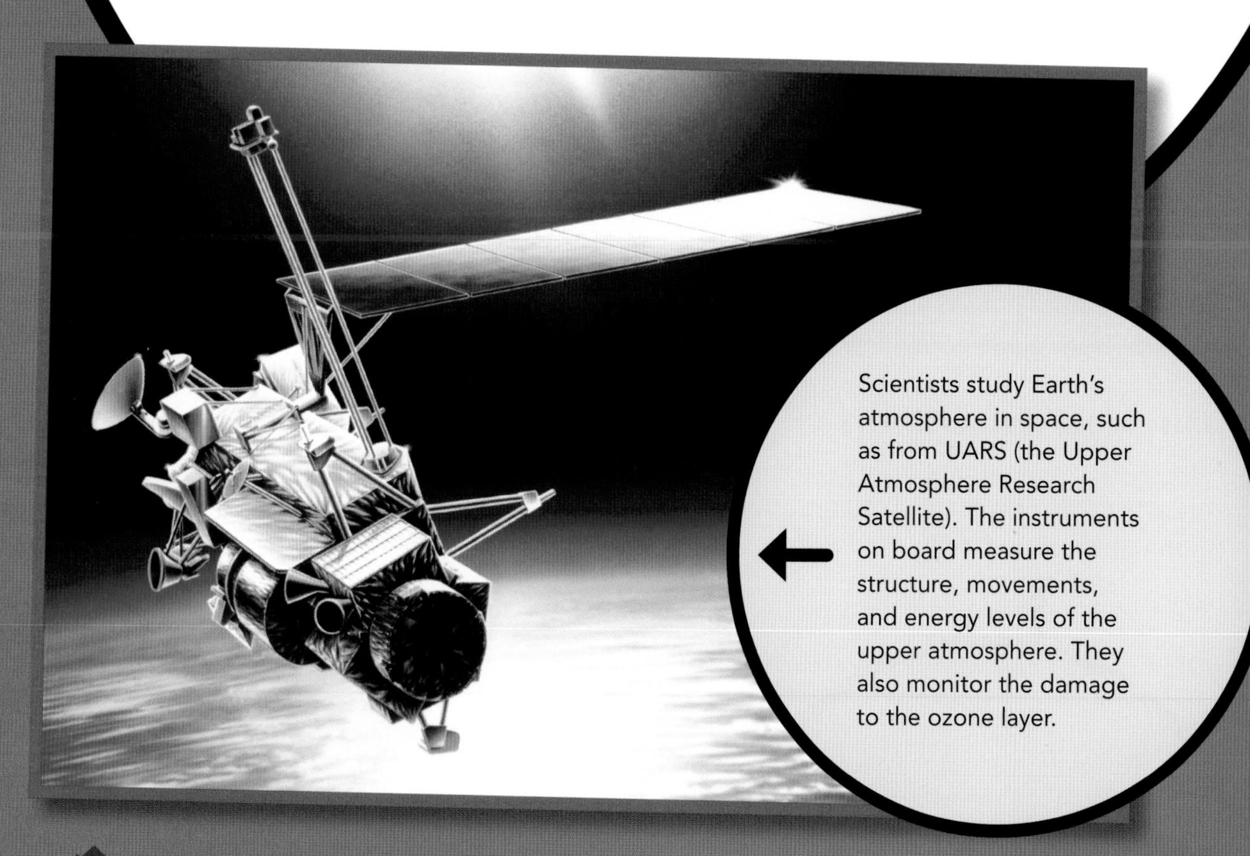

Scientists study Earth's atmosphere in space, such as from UARS (the Upper Atmosphere Research Satellite). The instruments on board measure the structure, movements, and energy levels of the upper atmosphere. They also monitor the damage to the ozone layer.

SCIENCE IN THE HOME: CFCs

Humans have been destroying the ozone layer by releasing greenhouse gases called CFCs (chlorofluorocarbons) into the atmosphere. These CFCs were used in our homes in spray cans and refrigerators. In 1987, many nations agreed to control the use of CFCs by replacing them with ozone-friendly gases. Many scientists now expect the ozone layer to recover. The hole in the ozone layer over the Antarctic could repair itself by 2065, but it may take hundreds of years for all the CFCs in the atmosphere to break down.

Carbon

Apart from changes to the ozone layer, Earth's atmosphere has remained much the same for millions of years. The balance of gases has been fairly constant, but the balance is now under threat from too much **carbon** dioxide.

All living things are made up of carbon. Carbon molecules are constantly changing and moving. If too much carbon is released into the air from the oceans, plants, animals, and rocks, the quantity of greenhouse gases warming Earth increases, as the carbon combines with oxygen to form carbon dioxide (CO_2). Over the last 100 years, the amount of carbon dioxide in the atmosphere has been steadily increasing.

Carbon dioxide

Every time we breathe out, we send carbon dioxide gas into the atmosphere. Every time we burn wood, coal, or oil, carbon is released and combines with oxygen to form carbon dioxide. Plants absorb carbon dioxide, which helps to stop a lot of it from entering the atmosphere. This movement of carbon dioxide from plants to air and back again is all part of the carbon cycle.

Back to the ocean

The oceans play an important part in Earth's carbon cycle. The total amount of carbon dioxide in the ocean is about 50 times greater than the amount in the atmosphere. Billions of microscopic ocean plants and animals absorb

The proportions of the main gases in the atmosphere are shown here. Nitrogen and oxygen together account for around 99 percent of the atmospheric gases. The remaining gases, such as carbon dioxide and water vapor, are present in far smaller amounts.

Gases in the air

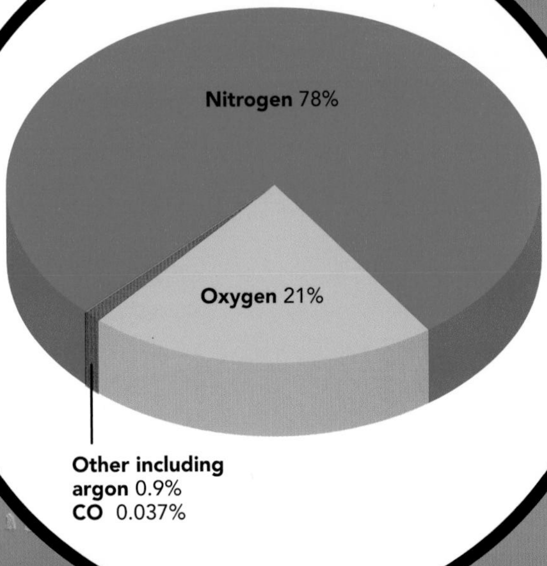

Nitrogen 78%

Oxygen 21%

Other including
argon 0.9%
CO 0.037%

vast amounts of carbon dioxide. This is known as a carbon sink. However, its future is in doubt if climate change affects ocean life, currents, and the water cycle, and reduces the number of microscopic plants in the oceans. The ocean will not be able to absorb as much carbon dioxide, so more will enter the atmosphere. The whole process could become not so much a carbon cycle as a vicious circle of climate change. At present, half the carbon dioxide produced by burning **fossil fuels** is absorbed into the oceans. The balance of the carbon cycle is being threatened like never before.

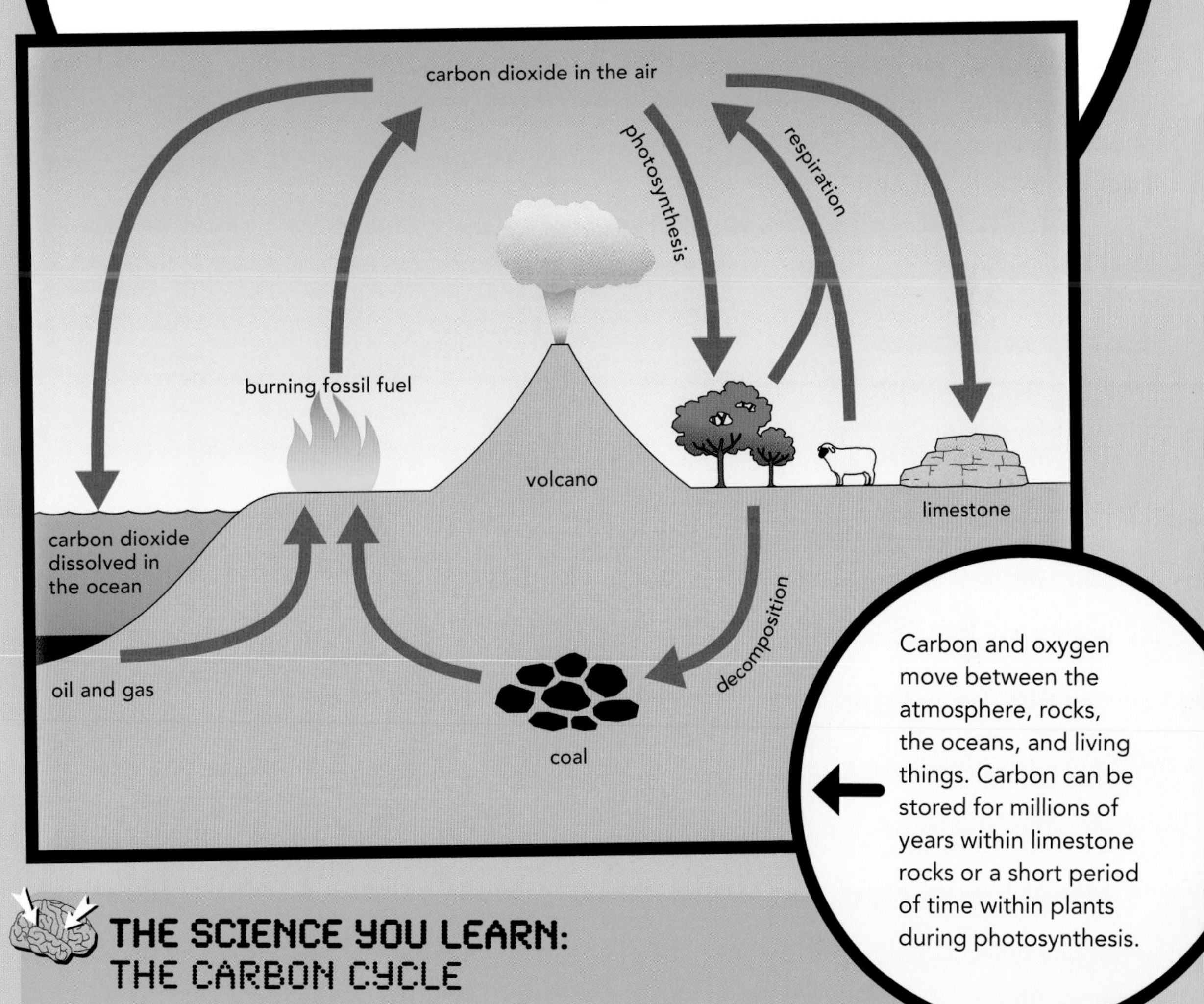

Carbon and oxygen move between the atmosphere, rocks, the oceans, and living things. Carbon can be stored for millions of years within limestone rocks or a short period of time within plants during photosynthesis.

THE SCIENCE YOU LEARN: THE CARBON CYCLE

Plants absorb carbon dioxide from the atmosphere or from water at the beginning of the carbon cycle. Some of this carbon becomes stored in the plant tissue and is passed on to plant-eating animals. Animals respire and return some carbon dioxide to the atmosphere when they breathe out. When animals and plants die, they decompose (break down). Carbon is released once more as carbon dioxide, to be used again by plants, and so the cycle continues.

Plants

Before looking at the evidence for climate change, it is worth remembering that the basic biology we learn is directly related to the delicate carbon balance on our planet.

More than 200 years ago, scientists discovered that plants take in carbon dioxide and release oxygen. This process is called **photosynthesis**, and it is driven by the Sun. Our knowledge of photosynthesis is vital in helping us understand Earth's carbon balance. If plants did not absorb carbon dioxide and release oxygen, life on Earth could not survive.

Although Joseph Priestley was not a trained scientist, he made some important discoveries, particularly about gases. At first he called carbon dioxide "heavy gas" because he noticed it was denser than air.

Early science

British scientist Joseph Priestley (1733–1804) discovered and experimented with carbon dioxide. In 1772, he found out how plants photosynthesize. He put a shoot from a green plant in a container, added a lighted candle, and sealed the container. The candle eventually went out when all the oxygen had been used up. Later, he discovered that the candle could be relit, which proved that the plant had produced oxygen when exposed to sunlight. He also dissolved carbon dioxide in water and invented carbonated water—he became the father of carbonated drinks!

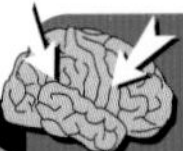

THE SCIENCE YOU LEARN: PHOTOSYNTHESIS

All life is affected by photosynthesis. In this process, plants use carbon dioxide and sunlight to make their own food in order to grow. They release oxygen, and the carbon becomes part of the plant. Vegetation has an important effect on the balance of gases on our planet—and therefore on Earth's climate. Photosynthesis can be written as an equation:

carbon dioxide + water —(sunlight and **chlorophyll**)→ glucose + oxygen

$$6\,CO_2 + 6\,H_2O \xrightarrow{\text{sunlight and chlorophyll}} C_6H_{12}O_6 + 6O_2$$

INVESTIGATION PANEL: OXYGEN PRODUCTION

A simple experiment can demonstrate how much oxygen is produced by a plant during photosynthesis. The best plant to use for this is a water plant such as pondweed.

Procedure

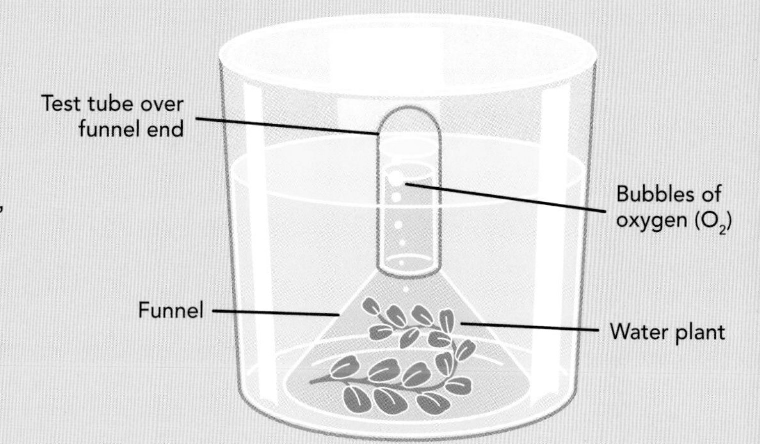

1. Immerse the plant in water in a large jar and cover the whole plant with an upturned, transparent funnel.
2. Place a test tube over the end of the funnel. Make sure there are no air bubbles in the tube.
3. Observe the apparatus and, after some time, bubbles of oxygen should rise up the funnel from the plant. They will eventually gather in the test tube and form a larger bubble.
4. You could also investigate whether the frequency or speed of the rising bubbles changes in bright sunlight. Can you work out what this proves?

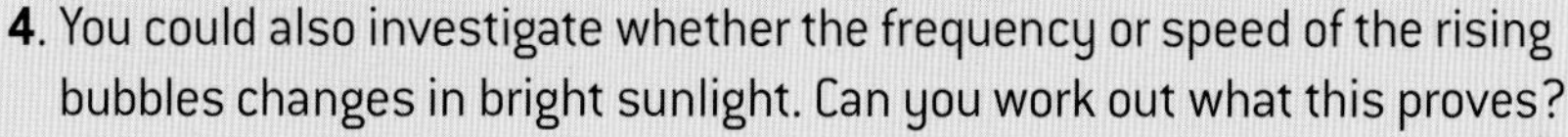

How does this science relate to climate change?

Humans influence the distribution of Earth's carbon dioxide in the carbon cycle. By destroying forests, we reduce the amount of oxygen produced and increase the levels of carbon dioxide in the air, because those forest plants soak up carbon dioxide. So fewer plants means an increase in atmospheric carbon dioxide levels, which may cause the planet to warm. Scientists believe that if the oceans warm, fewer ocean plants will be able to survive. At least half of the oxygen we breathe comes from the photosynthesis of ocean plants, so what will happen if climate change affects the oceans?

Evidence of Climate Change

One way to predict how today's climate change will affect Earth is to look for evidence of how Earth has been affected by climate change in the past. Conditions on our planet have rarely remained the same for more than a few thousand years at a time.

Changes in time

Geologists studying rocks and soil have discovered that many inland areas, including hills and mountains, were once under the sea. **Sedimentary** rock is made up of dead sea creatures that fell to the ocean floor hundreds of millions of years ago. Today, sedimentary rock is found far inland, which can sometimes demonstrate that sea levels have risen and fallen as the oceans have grown and shrunk throughout Earth's history.

Earth also has a long history of warming and cooling. We know that many parts of Earth that are warm today were once covered by ice. The movement of glaciers and massive ice sheets left behind deep scratches in rocks and many other clues that prove the climate was once far colder.

Geologists, biologists, and archaeologists dig down into the ground to learn about the past. They tunnel and drill through layers of soil, **sediment**, and rock. Each layer was formed at a different period in Earth's history, and the material they contain can provide useful information about the climate in that period.

Digging up the past

Scientists use special drills to take soil cores from the ground. A core is taken back to a laboratory where it is analyzed. The layers in the core can tell scientists a great deal about past climates. They can provide information about where glaciers have been, or where different plants once grew. For example, some plants only grow in certain conditions. If evidence of their existence is found in a soil core, they reveal what the climate must have been like at a specific time in the past.

Trees can provide a clear picture of any climate changes in recent years. It is possible to tell how old a tree is by counting the rings on a cross section of its trunk. A tree grows a new ring every year. The size of the tree ring tells scientists roughly how much rain fell in that year and the general temperatures.

When cores are taken from the ocean floor, ocean sediments provide a map of how ocean currents have flowed in the past. By taking many core samples from around the world, scientists have built up a picture of Earth's changing climate over thousands of years. But what exactly have they found out?

A soil core is a long cylinder of soil. The deeper parts are many years older than layers near the surface. Study of the types of pollen in various layers provides information about the changes in plant life and climate in the region over the last few thousand years.

Evidence in the ice

Some of the clearest evidence for the links among climate change, carbon dioxide levels, and human activity has come from ice. Just like soil cores taken from below ground, ice cores are drilled from the Arctic and Antarctic and are carefully stored in laboratories at below 0°C (32°F). Tiny bubbles of air trapped inside the ice from thousands of years ago can tell scientists about Earth's past climate. The air contains a record of the quantities of carbon dioxide, methane, oxygen, and nitrogen. In this way, scientists can study the rate of change of carbon dioxide levels. Readings of oxygen, and hydrogen from the water, give an indication of temperatures.

A scientist examines an ice core drilled from deep down in the Antarctic ice cap. This ice core came from a depth of more than 3 km (1.9 miles), and is around 900,000 years old. Analysis of the ice shows changes in climate during this time, and air trapped in the ice core tells us about the atmosphere when the snow fell. →

According to the U.S. Environmental Protection Agency (EPA), during the last 650,000 years, carbon dioxide levels in Earth's atmosphere have gone up and down at the same time as rising or falling temperatures. During warm periods carbon dioxide levels have been high, and during cool, glacial periods carbon dioxide levels have been low. Ice studies show that levels of carbon dioxide in the atmosphere are now higher than they have been for more than half a million years. This is having a profound effect on Earth's climate.

In the distant past, the rise in carbon dioxide levels has slightly followed, or "lagged," increases in temperature. Past rises in temperature were not caused by human activity. But scientists now know that the way carbon dioxide levels rose is complicated. They know that the more human activity increases carbon dioxide, the even greater the effect on temperature.

Other evidence

- Arctic temperatures have increased since the 1980s by up to 3°C (5°F). In the Russian Arctic, buildings are collapsing because ice under their foundations has melted.
- Snow cover has declined by 10 percent in many places since the 1960s.
- The overall volume of glaciers in Switzerland decreased by about 66 percent in the 20th century.

CASE STUDY

Data from Buried Mud

Jason Briner is a geologist from the State University of New York in Buffalo. He and his team are gathering temperature data from the last thousand years on Baffin Island. To do so they are studying mud buried many feet below the surface of lakes in the frozen Canadian Arctic.

Using lake sediments from Baffin Island, Jason Briner has shown that this area of the Arctic has experienced temperatures about five degrees warmer than today. Clues in the cores include fossils and organic material from dead organisms and **algae**. "Generally, the more organic matter in sediments, the warmer the climate," says Briner. "In the past 100 years, both the magnitude and the rate of temperature increase exceed all the variations of the past 1,000 years."

These researchers are collecting an ice core from an ice floe in the Weddell Sea.

More scientists at work

Many scientists are studying the world's ice to measure how it has melted in the past, the rate at which it is melting now, and whether it will melt at greater rates in the future. Not all research involves studying the ice through a microscope. One way to examine the changing state of the Arctic's ice is from space. Arctic sea ice has been found to be shrinking faster now than ever before. This is having a major effect on the planet and will continue to do so.

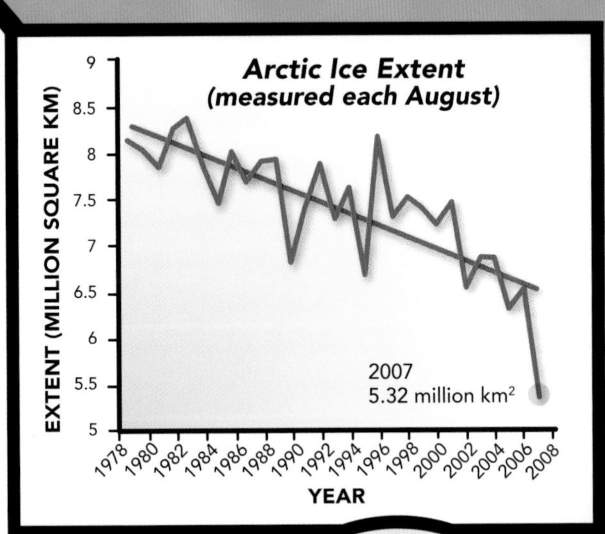

This graph shows that Arctic ice is shrinking rapidly. The red line shows the trend of the decline.

CASE STUDY

NASA

The whole of planet Earth is scanned every day from space. **NASA's** Earth Observing Fleet orbits the planet more than 10 times each day and measures changes on Earth. The Aqua satellite first went into space in 2002. It measures microwaves and gives precise and very detailed measurements about snow and ice cover. The ICESat (Ice, Cloud, and land Elevation Satellite) spacecraft was launched the following year and measures changing surface features of worldwide ice cover. It pulses a laser beam at Earth 40 times per second and measures the reflected light, allowing scientists to track changes in the Arctic and elsewhere. Another NASA satellite measures thermal emissions (temperature) from the planet. All of this data helps scientists clearly see any changes in the Arctic, and provides information about the possible effects on climates worldwide.

The Arctic Warming Study has shown that most of the Arctic has warmed over the last decade, with the biggest temperature increases occurring over North America. Sea ice in the Arctic is shrinking at a rate of 9 percent every 10 years, and in 2002 summer sea ice was at record low levels.

The Antarctic

Ice is not only melting at the Arctic. The United States Space Agency's research into climate change has also shown a dramatic thinning of Antarctica's ice. Some glaciers are disappearing into the ocean eight times faster than they were 10 years ago.

Information can be gathered by low-level aircraft fitted with special **radar** equipment. The radar measures the thickness of the ice, while laser equipment measures the height of the ice above the ocean. Surveys over recent years have shown that Antarctica has been losing far more ice to the sea than it is gaining from snowfall. The melted ice has entered the oceans and caused sea levels to rise. Even if the rise in sea level stays at about its current rate of 2 mm (0.08 in.) per year, the consequences will be serious for many areas of the world. Any rise in sea level greater than this could cause major problems all around the world.

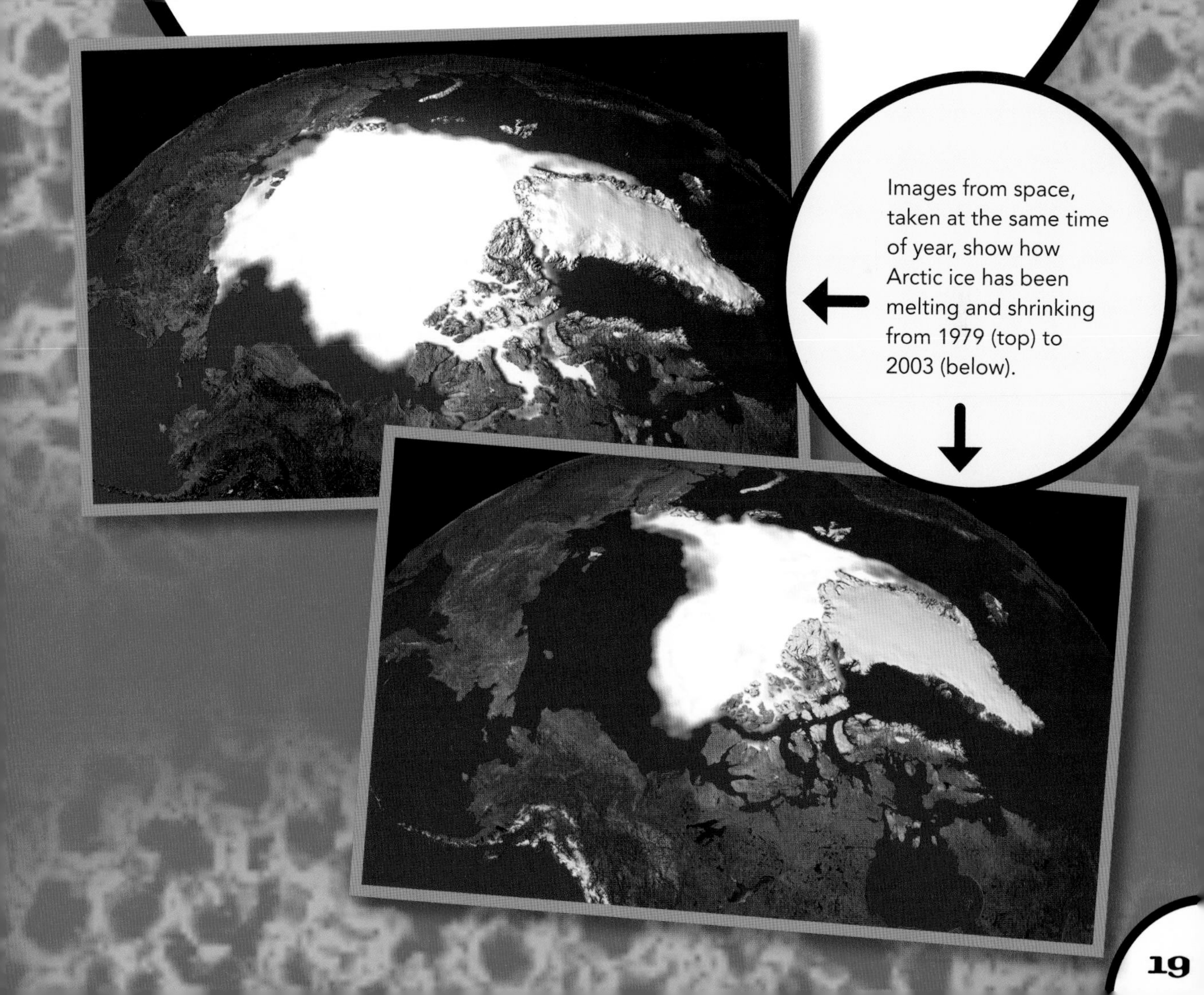

Images from space, taken at the same time of year, show how Arctic ice has been melting and shrinking from 1979 (top) to 2003 (below).

Further research

Scientists continue to study the oceans to gather further evidence of climate change in the hope of predicting what might happen next. About one-quarter of the 30 billion metric tons (33 billion tons) of carbon dioxide produced by humans each year is thought to be absorbed by the oceans. However, the rate at which the oceans absorb and release carbon dioxide seems to vary every year, and scientists want to understand why. "Understanding how carbon dioxide reacts with cold surface water is important for determining how the ocean uptake of carbon dioxide will respond to future climate change," said Christopher Sabine, an American scientist on the 2008 Southern Ocean Gas Exchange Experiment, or GasEx III.

This diver is checking a net used to collect plankton. Plankton can provide information on the health of a marine ecosystem, which can be affected by climate change.

The evidence so far

Results from scientists' research so far all point to the same conclusions.

- Earth is warming. Eleven of the last twelve years rank among the warmest years in global surface temperature since 1850. The average global temperature increased by about 0.74°C (1.3°F) during the 20th century. The warming has affected the land more than the oceans.
- There is now more carbon dioxide in the atmosphere. The amount of carbon dioxide in the atmosphere was around 280 ppm (parts per million) in 1750. Today, carbon dioxide levels have reached over 385 ppm.
- The oceans are changing. The average temperature of the world's oceans has increased in depths of at least 3,000 m (9,840 ft.). The total 20th-century rise in sea level is estimated to be 17 cm. Scientists believe that sea levels have risen partly because glaciers and sea ice have melted and released their water into the oceans. In addition, as heat is added to water, it expands. As the oceans have warmed, they have expanded and taken up more space.
- There is now less snow. Snow cover is decreasing in most regions. Since 1900, the amount of frozen ground has decreased by 7 percent in the northern hemisphere.
- Ice is melting. Mountain glaciers and snow are melting and have contributed to a rise in sea level of 0.77 mm per year between 1993 and 2003. The shrinking of Greenland's and Antarctica's ice sheets have contributed to a rise in sea level of 0.4 mm per year between 1993 and 2003.

All of this evidence provides clear proof of climate change.

The Industrial Revolution began around 1750. Factories began pumping smoke into the atmosphere. The amount of carbon dioxide in the atmosphere was around 280 ppm (parts per million). Today, information from ocean scientists, data from ice cores, and measurements from satellites and weather stations build a clear picture of increasing atmospheric carbon dioxide levels. The amount of carbon dioxide in the atmosphere has rocketed to over 385 ppm.

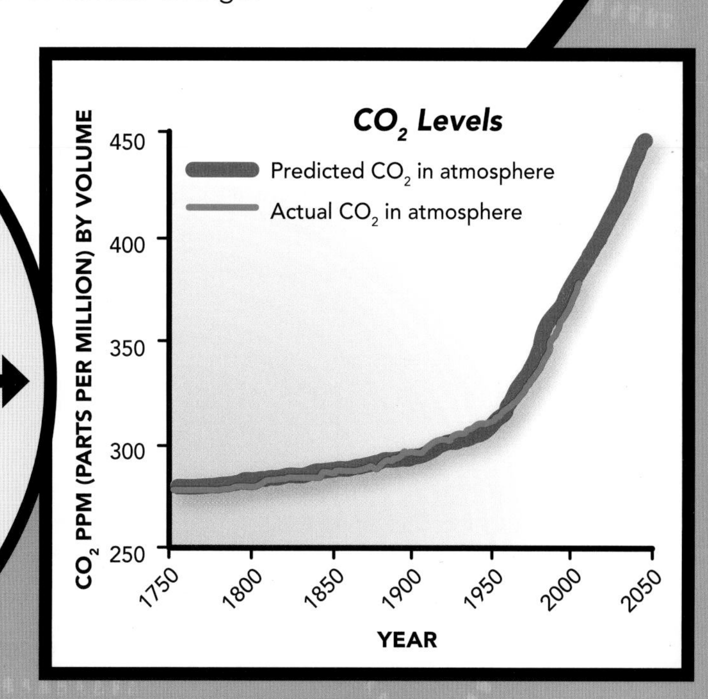

The Earth Simulator (ES) was the fastest supercomputer in the world between 2002 and 2004. The system analyzes and predicts environmental changes through the **simulation** of global warming, air and sea pollution, flooding, and other complicated environmental effects.

Using the evidence

Gathering climate change evidence together, making sense of it, and trying to predict what will happen next is one of the biggest challenges facing scientists. Computers play an extremely important role in trying to predict the future of our planet.

The problem is that however powerful the computer and its software, climate change prediction is still an inexact science. Even the most advanced computers can be incorrect if inaccurate data is entered, or if just one detail is overlooked. It is difficult to be 100 percent accurate predicting this type of information. After all, daily weather forecasters' computers have been known to get things wrong!

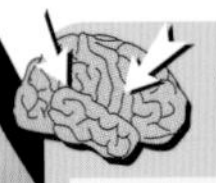

THE SCIENCE YOU LEARN: HOW SCIENTISTS WORK

Scientists develop a **hypothesis** and test it through investigation. They gather data (evidence) to support or disprove the hypothesis. Sometimes an interesting set of new data can trigger a new hypothesis that then needs to be investigated. Sometimes there can be conflicting evidence that needs to be understood.

With climate change, there are many factors to consider because the whole climate system is so complex.

Many science organizations predict climate change using complex computer programs to calculate how different parts of the climate jigsaw puzzle fit together. A huge supercomputer in Japan, called the Earth Simulator, processes masses of data from weather stations, satellites, and ocean scientists around the world. It calculates millions of equations at great speed, using data records of temperatures, rainfall, carbon dioxide levels, and sea levels. It shows how each of these factors affects the others. All together, the equations provide a picture of how the climate is working and developing.

Uncertain future

Even huge computers dealing with millions of calculations have their limitations. After all, just a single event could suddenly make predictions out of date. For example, a massive forest fire or a major volcanic eruption could quickly affect the atmosphere and alter the whole balance of the equations. Scientists have to build all kinds of **variables** into their models, such as how much sunlight is reflected and absorbed by Earth's atmosphere, the temperature of the air and oceans, the distribution of clouds, rainfall, and snow, and what may happen to the polar ice caps. They have to consider all the likely events and resulting effects that might change the developing picture.

Another problem with computer models is that they base their predictions on what has already happened, and then project the trends into the future. This assumes that the variables do not change too much. However, in the real world, many natural and human activities can trigger changes. This is why the farther into the future these computer models run, the less accurate they are likely to be.

NASA's global climate computer model (GCM) shows how Earth's climate is changing. It calculates how much sunlight is reflected and absorbed by Earth's atmosphere as well as the temperature of the air and oceans. Areas colored red show the biggest predicted increase in surface air temperature by 2040.

Causes of Climate Change

With so many scientists and politicians trying to address climate change, there are heated arguments about what is causing it. Unless we address the causes, how can we stop it from worsening? Although increased levels of carbon dioxide and other greenhouse gases are generally agreed to be a major cause, are humans or nature to blame?

Natural causes

Volcanoes can dramatically affect weather conditions. When a volcano erupts, it throws out masses of sulphur dioxide, water vapor, carbon dioxide, dust, and ash. Millions of tons of debris can reach the upper atmosphere. The gases and dust block incoming sunlight, causing cooling. Sulphur dioxide combines with water to form tiny droplets of sulphuric acid. These droplets are so small that many stay in the atmosphere for years. They reflect sunlight away from Earth back into space, and screen us from some solar energy.

If volcanoes have a cooling effect, how can they be blamed for global warming? Scientists are measuring how volcanoes contribute to the greenhouse effect.

Winds in the upper atmosphere carry the **aerosols** (the mixture of particles and gas) around Earth, and cooling can last for years after a major volcanic eruption. In 1815, Tambora, an Indonesian volcano, caused "the year without a summer." There were major weather changes in North America and Western Europe, including summer frosts in the United States and Canada.

In 1991, Mount Pinatubo in the Philippines erupted, producing the largest sulphur dioxide cloud of the century. The combined cloud of Pinatubo and an eruption by Mount Hudson in Chile spread around Earth in months. Data later showed that the cooling effect of the sulphur dioxide lowered world temperatures by about 0.4°C (0.7°F) over the next two years.

CASE STUDY

Volcanic Activity and Climate Change

In 2007, British scientists began to study an active volcano on the island of Montserrat in the Caribbean. They hope to find out what effects volcanic activity has on the oceans and on climate change.

"Volcanoes can influence the climate as they emit a wide variety of gases including carbon dioxide, which we know is a greenhouse gas, and also deliver materials to the sea floor and oceans," said Martin Palmer, a geography scientist at the National Oceanography Center.

Volcanoes are known to suck up oxygen from sea water, disturbing the ocean's oxygen and carbon balance. Over 10 years, the Montserrat volcano has deposited about 90 percent of its erupted materials into the sea. The study will look at what happens to life on the ocean floor around the volcano by studying and analyzing sediments. No one has studied this before. The scientists then plan to return in a few years for deep-sea drilling around the island. This will provide them with a detailed history of the area.

The United States produces enough carbon dioxide in one year to cover the entire country with 30 cm (12 in.) of carbon dioxide. With many other countries adding to the estimated production of 6,440 million metric tons (7,100 million tons) per year of carbon dioxide, there is bound to be a worsening of the greenhouse effect.

Human causes

Nature is so powerful, and the scale of Earth's weather systems is so vast—can humans really affect the planet's climate? Human activity has altered Earth's atmosphere in two main ways.

1. We have cleared vast areas of natural vegetation, for building and agriculture, which affects Earth's balance of oxygen and carbon dioxide.
2. We have greatly increased the amount of greenhouse gases in the atmosphere. Every time we use electricity, heat our homes, or travel by car or plane, we produce carbon dioxide. Agriculture increases the amount of other greenhouse gases.

Agriculture

A major contributor to greenhouse gas emissions and the resulting climate change is farming. Farmers use fertilizers to grow high-yielding crops. Fertilizer production and use is causing levels of the greenhouse gas nitrous oxide to increase.

Methane, another greenhouse gas, is also on the increase. About 25 percent of all methane gas in the atmosphere is thought to come from farm animals such as cows, goats, pigs, horses, and sheep. Animals produce methane as a result of bacteria in their stomachs. Methane is also released from paddy fields (flooded fields) used to grow rice. Bacteria in flooded soil produces methane. As a result, a lot of methane gas is released into the atmosphere all across Asia, where rice is widely grown.

In many parts of the world, forests are cleared to make space for agriculture. Every year, around 13 million hectares (32.1 million acres) of tropical forest—an area larger than Austria—are cleared. The fires used to clear the forests release masses of carbon dioxide, and fewer trees means less carbon dioxide is absorbed from the atmosphere.

DID YOU KNOW: COW GAS

Every day a cow burps a lot of methane gas—about 280 liters (74 gallons)! The average cow could fill 140 2-liter bottles with gas in one day. Around 6 million metric tons (6.6 million tons) of methane float up into the atmosphere every year from herds in the United States alone. Cattle around the world probably create 80 million metric tons (88.1 million tons) of methane per year.

In 2007, forest fires in the United States pumped as much carbon dioxide into the atmosphere in a few weeks as all the motor traffic in the country produces in a year, according to scientists at the National Center for Atmospheric Research (NCAR).

The research continues

The science of climate change is complex and sometimes uncertain. While scientists agree that Earth's climate is changing quickly due to increasing concentrations of greenhouse gases caused by human activity, there is still a great deal of detail to understand.

Here are just three of the many topics that scientists are researching:

1. The nature of methane gas emissions. Farming certainly produces much of the increasing emissions, but plants are now also thought to be a major source. An international team of scientists estimates that the world's plants produce more than 150 million metric tons (165 million tons) of methane each year. That's 20 percent of all the methane that enters the atmosphere. Future studies will measure how much of an impact plant-produced methane actually has on the environment.
2. Scientists know that computer models of climate change can be improved. They know that simulating the complex nature of our climate systems is difficult, even using the most powerful supercomputers. Cloud formation, ocean temperatures, and the mixing of the air are still not fully understood, and computer climate models do not all agree in their predictions. Despite the uncertainties, American NASA climate modeler Drew Shindell says: "Climate models are getting better. Many climate models run by scientists around the world describe a world sweating under the influence of greenhouse gases which trap the Sun's energy. With the same amount of energy going in and less going out, the Earth has to warm up. That's elementary physics."
3. Some scientists are studying the impact of natural climate change. They are studying how much influence the Sun's activity is having on climate change compared to the rising load of human-produced, heat-trapping gases in the atmosphere.

However much scientists continue to study the finer details of climate change, they agree that the world will continue to experience its effects for the next 100 years. Many of those effects are being studied by scientists across the world as they try to work out the impact of climate change on our lives.

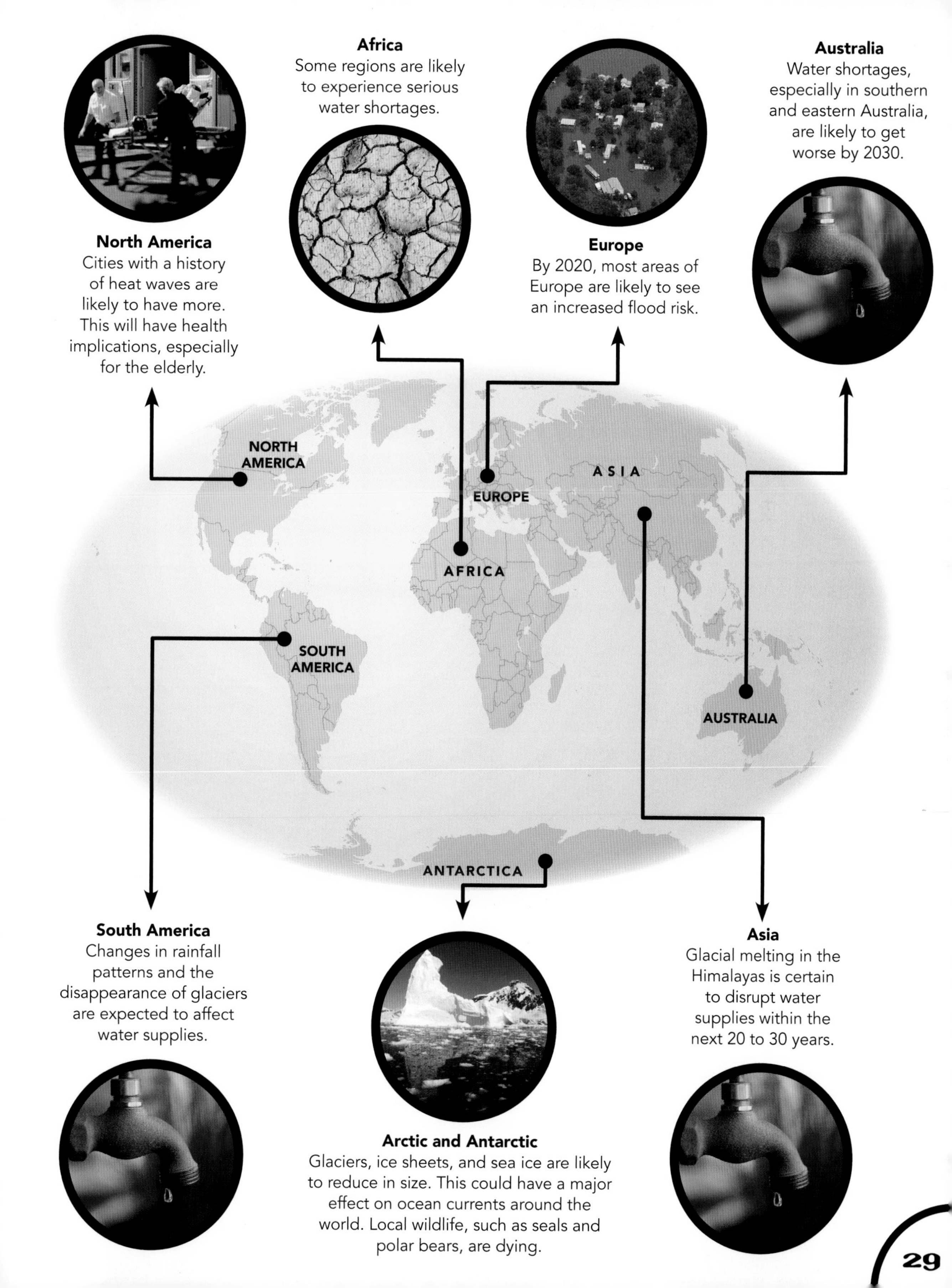
North America
Cities with a history of heat waves are likely to have more. This will have health implications, especially for the elderly.
Africa
Some regions are likely to experience serious water shortages.
Europe
By 2020, most areas of Europe are likely to see an increased flood risk.
Australia
Water shortages, especially in southern and eastern Australia, are likely to get worse by 2030.
NORTH AMERICA
EUROPE
ASIA
AFRICA
SOUTH AMERICA
AUSTRALIA
ANTARCTICA
South America
Changes in rainfall patterns and the disappearance of glaciers are expected to affect water supplies.
Arctic and Antarctic
Glaciers, ice sheets, and sea ice are likely to reduce in size. This could have a major effect on ocean currents around the world. Local wildlife, such as seals and polar bears, are dying.
Asia
Glacial melting in the Himalayas is certain to disrupt water supplies within the next 20 to 30 years.

Effects of Climate Change

Climate change is already melting ice, raising sea levels, causing floods and droughts, and threatening plant and animal species. Scientists predict that these problems will continue, but for how long, at what rate, and what might be the long-term effects?

Meltdown

Does it matter if the world's ice is gradually melting? In fact, some people may see a benefit in having less ice. Ships will be able to travel more easily through sea routes that are currently frozen solid in winter. People living near the Arctic and Antarctic will save on their heating costs. However, the melting water has to go somewhere. There is so much ice at the Arctic and Antarctic that if it all melted, global sea levels could rise as much as 80 m (262 ft.). Most of the world's major cities and ports, such as London, New York, and Sydney, would be under water.

Argentina's Upsala Glacier was once the biggest in South America, but it is now disappearing at a rate of 200 m (655 ft.) per year.

CASE STUDY

The End for Polar Bears?

A study by scientists from the Canadian Wildlife Service discovered that the sea-ice season in Hudson Bay has been reduced by about three weeks over the last 20 years. Temperatures in the Canadian Arctic have risen by 4°C (7°F) in the last 50 years, and polar bear numbers have dropped by nearly 25 percent in just 20 years.

Melting snow and ice means that the polar bears' habitat is disappearing. Polar bears dig dens in snowy mountain slopes or in sea ice and use these homes to shelter with their young. Polar bears also need sea ice in order to hunt seals. During the winter, the polar bears hunt, eat, and build up their body fat. After the winter, they come ashore and rely on this fat to survive and to feed their cubs. The shrinking ice is reducing their hunting grounds. Climate change is effectively causing polar bears to starve.

Increasing numbers of hungry bears now wander into the northern Canadian community of Churchill, Manitoba, looking for food. Many of the surviving bears are very thin. When polar bears lose their layers of fat, they can die from the cold. Mothers now tend to have only one cub rather than several young at a time. Scientists are worried that all of these factors could lead to the extinction of wild polar bears.

Seals are the main prey of polar bears. Arctic warming melts the snow caves where female seals give birth. Their young have no blubber and die without protection from cold winds. Scientists think the decline in seal numbers may be partly causing the decline in polar bear numbers.

In the United States in 1998, parts of Texas had 29 consecutive days with temperatures above 38°C (100°F). Scientists are expecting the southwest United States to dry up during this century.

Global warming

Some scientists believe that by the time babies born today reach 80 years old, the world will be 3.5°C (6.3°F) warmer. Others predict it will be even warmer! At first, this might seem welcome. Hotter summers and warmer winters could be very pleasant. Many areas would have frost-free winters and all kinds of new plants could be grown. However, there will be another side to global warming—heat waves. Very high summer temperatures are not a good thing.

CASE STUDY

Predicting Heat Waves

Between 1979 and 1999, there were 8,015 heat-related deaths in the United States. Illness and deaths increase during heat waves, such as the 119 deaths recorded in Chicago during one heat wave in 1999. Scientists are using computer models to predict future heat waves. They predict that throughout much of the Midwest of the United States, during the worst heat waves, nighttime temperatures will be 2°C (3.5°F) higher than in previous heat waves. Scientists are also predicting a 70 percent increase in the number of heat-wave days per year for the Midwestern region by the late 21st century. This is bad news for people with health problems.

During the summer of 1998, parts of the United States suffered record high temperatures of up to 46°C (115°F) and more than 140 people died. In 2003, nearly 15,000 people died in a blistering heat wave in France, when temperatures reached 42°C (108°F) for many days in a row. Heat and drought can cause widespread crop failures, with stunted corn producing less than half its normal yield, causing serious food shortages.

Why warming up can be unhealthy

The World Health Organization estimates that already 150,000 people are killed by climate-change-related problems every year. Famine is one of the main problems. Poor areas of the world are most at risk, particularly where deserts spread, droughts increase, and crops fail. Richer countries face health problems, too.

All countries are affected. Authorities in China calculated that between 173 and 685 Chinese citizens per million die every year from sickness related to global warming, such as strokes and heart disease. Doctors warn that global warming will cause more heart disease: "The hardening of the heart's arteries is like rust developing on a car," said Dr. Gordon Tomaselli, a heart specialist at Johns Hopkins University. He said that just as rust develops much more quickly at warm temperatures, so too does heart disease.

Will city-dwellers have to wear face masks for protection against smog as temperatures keep rising?

Breathing in **smog** caused by vehicles in urban areas can be very dangerous as well. Warm air is a key ingredient in smog, so warmer temperatures will likely increase smog levels. Canadian doctors say smog-related deaths could rise by 80 percent over the next 20 years.

Sea change

Millions of people around the world live near the seas and oceans. Climate change is threatening them like never before. Many scientists believe that sea levels will be about 1 m (3.3 ft.) higher than they are now by the end of this century. The effects could be catastrophic, with local flooding and storm surges sweeping homes away. Current flood defenses along our coasts would not be able to cope with the rising water.

Not all sea levels are expected to rise over the next 100 years. Some seas and lakes could dry up entirely. West Africa's Lake Chad has shrunk to 5 percent of its former size. Central Asia's Aral Sea is shrinking and gradually turning into desert. More than half of the world's 5 million lakes are endangered. This means a loss of freshwater and a blow to fishing and other industries.

Rainfall and storms

Global warming is also changing rainfall patterns and causing more intense rainstorms, across parts of the northern hemisphere, for example. The intensity of all sorts of storms will increase. In June of 2008, continuous heavy rains flooded areas along rivers in the Midwest, causing thousands of people to flee their homes. Many areas had received as much rain by the end of July as they might normally get in an entire year.

Scientists predict that a warmer climate will trigger more violent storms and increased coastal erosion, particularly as sea levels keep rising. This lighthouse on the coast at Cape Hatteras in North Carolina had to be moved 884 m (2,900 ft.) in 1999 to keep it from falling into the Atlantic Ocean. The old site of the lighthouse is marked by a circle of stones in the foreground.

Vanishing lakes and rivers

The effects of climate change on Earth's water could well be the biggest threat to humankind. The Ganges River in India is beginning to run dry. The river is fed by a glacier in the mountains, which is shrinking at a rate of 40 m (130 ft.) per year, which is nearly twice as fast as 20 years ago. Scientists warn that the glacier could be gone by as soon as 2030. The lack of water will destroy the livelihoods of millions of people who depend on the river for farming and fishing. Wildlife will suffer, too, as habitats are devastated.

African rivers are also drying up. Scientists predict a 20 percent drop in rainfall in parts of Africa by 2070. That would leave many rivers almost dry at various times of the year, which would be devastating for local farmers and villages.

Scientists in Chile have blamed climate change for the disappearance of a lake in the south of the country. At the beginning of 2007, the lake was there, but a few months later it had gone. All that was left was a huge, dry crater and stranded chunks of ice that had previously floated on the water. A glacier that acted like a dam holding back the lake had melted and allowed the lake to drain away.

Climate change will bring more hurricanes and storms to our coasts. Will images like this become more common?

Bleak biology

Biologists are warning that climate change will damage the balance of life on the planet forever. Ecosystems will come under threat as plants, animals, and **food chains** adapt or disappear.

Scientists at Harvard University predict that climatic changes will alter the health of humans and ecosystems worldwide. They say that infectious diseases, extreme weather events, and ecosystems such as forests and ocean habitats will be affected. They also warn that rising levels of carbon dioxide may be increasing pollen production by plants, resulting in an increase in the incidence of asthma. In fact, the World Health Organization has warned of more than 30 human diseases on the increase because the world is getting warmer.

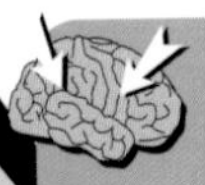

THE SCIENCE YOU LEARN: MOSQUITOES AND MALARIA

Living things within a habitat influence each other and are affected by their environment. This is certainly true of mosquitoes. There are about 3,500 species of mosquito, but it is the female *Anopheles* mosquito that is dangerous to humans. It carries the deadly disease, malaria. These mosquitoes have adapted to live in warm, damp habitats. Malaria is therefore a serious problem in tropical environments. Climate change is expanding the mosquitoes' habitat. Mosquitoes can now thrive in habitats that were once too cold.

Already, warmer temperatures have caused malaria to spread from 3 districts in western Kenya to 13, and have led to epidemics of the disease in Rwanda and Tanzania. Malaria-carrying mosquitoes can live further north outside the tropics, which could lead to a surge in malaria across Europe. Malaria has also spread to higher altitudes that were once too cold, such as the Colombian Andes, 2,135 m (7,000 ft.) above sea level.

Climate change may be good news for mosquitoes, but bad news for people around the world.

Animals in danger

The latest report from the World Conservation Union says that 40 percent of the world's animal species are being threatened because of global warming. The following are just some of the species in danger.

- Koala bears. Australia's Climate Action Network has reported that higher temperatures are killing off eucalyptus trees, the koala bears' main food. Without eucalyptus leaves, these bears could become extinct in the next few decades.
- Frogs. An estimated 66 percent of the 110 known species of harlequin frog in Central and South America have vanished since the 1980s due to the outbreak of a deadly frog fungus. The fungus is believed to have spread as a result of global warming. According to one scientist: "Disease is the bullet that kills the frogs, but climate change is pulling the trigger."
- Arctic foxes. The white Arctic fox is disappearing as temperatures rise. The more aggressive red fox is moving north and taking over.
- Snails. The Aldabra banded snail is now extinct. It existed only on one island at the northern tip of Madagascar. The snail died out when warmer weather reduced the rainfall in its habitat.

Coral reefs are under greater threat than ever before. Experts warn that the Great Barrier Reef in Australia will be dead within 30 years if urgent action is not taken to prevent the effects of climate change and pollution.

What about daily life?

Climate change could mean disruption on a huge scale for life on the planet, but what about daily life for the average person? Will climate change really impact most people's lives?

Some people might not worry about climate change in faraway places. However, they might think again if their own food is affected.

- Salmon. Wild Pacific salmon have already vanished from 40 percent of their habitats. They are cold-water fish and warmer temperatures will only reduce their habitats further.
- Lobsters. Lobsters thrive in the cold waters of New England, but their numbers are falling as the waters have warmed up.
- Avocados and nuts. Scientists from the Lawrence Livermore National Laboratory in California predict that hotter temperatures will cause a 40 percent drop in California's avocado production over the next 40 years. They also predict a 20 percent drop in California's almond and walnut crops.
- Chips. Scientists from the Consultative Group on International Agricultural Research say warmer temperatures are killing off species of potato plants.

The Tuvalu Islands in the South Pacific Ocean are struggling to survive against rising sea levels. During a "king tide," all the lowest areas of the islands are flooded. Many islanders are unable to grow food crops because the land has become too salty.

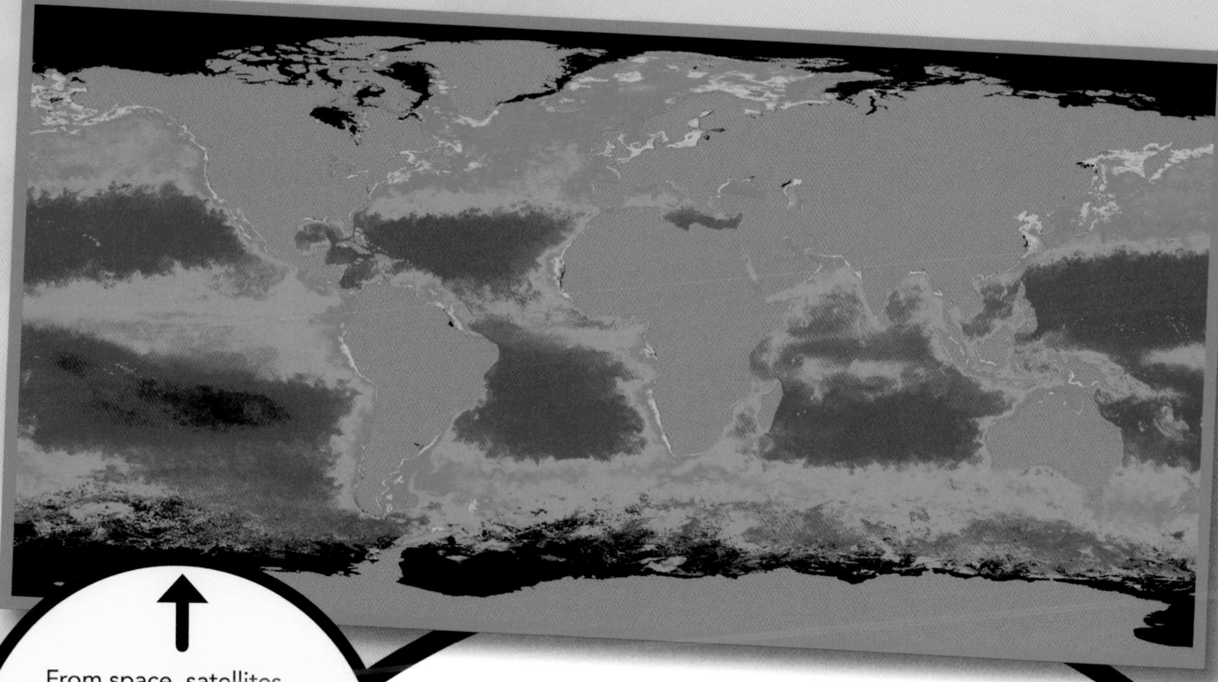

From space, satellites measure ocean plant productivity. Here, high plant productivity is shown in green, while areas of low productivity remain blue.

What more could happen?

Alarming new satellite data shows that the warming of the world's oceans is reducing marine life. The ocean's food chain is based on billions of **phytoplankton**, which are now disappearing. This is threatening ocean fisheries like never before.

The effects of climate change will be even wider. As there is less rainfall and rivers dry up on the Indian subcontinent and Africa, there will be less food and water to support the population. This will result in the mass movement of people trying to find the resources they need to stay alive. People and governments will not be able to keep them from moving. They will become refugees, needing aid from other people. But many people will not want to give money and other aid. They will want to protect their own standard of living. With fewer people able to look after themselves, growing food, and earning money, the world's economy will be affected. There is a huge risk of political conflict and wars breaking out as economies and governments are unable to support their people.

Maybe when people become hungry they will consider the big question facing us all. What can *we* do?

What Can We Do?

The world must act fast. Already governments are making important decisions to help cut down greenhouse gas emissions. Many world leaders see climate change as one of the biggest challenges facing us all. The world is now listening to scientists, but it is up to everyone to play their part in tackling the problem.

World response

In 2007, scientists reported that climate change is "a wake-up call for world leaders. Even if carbon dioxide emissions are slowed or stopped, temperatures and sea levels will continue to rise, low-lying lands will be flooded, and many life forms will become extinct." The **United Nations** Secretary responded with "Global warming is the defining challenge for our age."

Tackling climate change

The United Nations holds regular meetings in which world leaders discuss how to tackle climate change. Every five years they meet at a conference called an Earth Summit. In 1997, Earth Summit+5 was held in New York, and a Climate Change Conference took place in Kyoto, Japan. The aim of these conferences was to form important agreements among nations to cut the greenhouse gases that industries produce. Most industrialized nations agreed to reduce greenhouse gas emissions by 5 to 8 percent below 1990 levels, between 2008 and 2012.

Another conference occurred in Bali, Indonesia, at the end of 2007. More than 180 countries attended. Despite difficult negotiations, further important agreements were made, including the Bali Roadmap Agreement to move toward an agreement to replace the Kyoto provisions in 2012. Dutch Secretary Yvo de Boer said: "This is a real breakthrough, a real opportunity for the international community to successfully fight climate change."

Different countries, different solutions

Many politicians think that developed countries have the greatest responsibility to stop carbon emissions because they produced the most carbon in the past. In rapidly developing countries, such as China and India, greenhouse gas emissions are increasing at an alarming rate, and many people feel that developed countries have a responsibility to help them. This means assisting with **renewable energy** technology such as solar, wind, water, wave, tide, and **geothermal** power.

Renewable energy currently provides 17 percent of world energy needs. Countries developing renewable energies include Brazil (particularly **biofuels**), China (solar hot water and hydropower), Denmark (wind power), Germany (wind and solar power), India (wind, solar power, and biofuel), Japan (solar power), and Spain (wind power).

Since 1990, some developed countries with large industries have already reduced greenhouse gas emissions by using renewable energy and low-emission technologies, and by replanting forests. Norway has reduced emissions by 18 percent, Germany by 17 percent.

More than 860 **Clean Development Mechanism projects** in 49 developing countries, with many more planned, are expected to reduce carbon emissions by billions of tons by 2012.

The Bali Roadmap for an international agreement on climate change was adopted at a United Nations Climate Change conference in 2007.

Engine technology

Engineers are developing new ways to lower carbon emissions. Driving a car makes up about 40 percent of the average person's greenhouse gas emissions. Biofuel cars, such as those in Saab's 9-5 range, run almost entirely on fuel made from wood chips, wheat, and sugar. These plants absorb carbon dioxide as they grow, helping to keep the carbon cycle balanced. Such cars are already popular in Brazil and Sweden. **Hybrid** cars, such as the Toyota Prius, burn less fuel by having gasoline engines and electric motors. Hybrids and biofuel cars help to cut carbon emissions, but are there cars that produce no greenhouse gas emissions at all?

David Hart, a scientist at Imperial College London, works on fuel-cell vehicles. The fuel is hydrogen gas, and the vehicles emit hardly any carbon. They work by converting chemical energy into electrical energy, like a battery. Their main emission is water. David Hart believes that if everyone used hydrogen fuel-cell cars, carbon dioxide emissions from vehicles could drop to almost zero by 2050. One problem with this technology, however, is the way the hydrogen is obtained in the first place, using electricity from fossil fuels. For these vehicles to be truly nonpolluting, the hydrogen must be produced using a renewable, clean energy source, such as solar energy.

This hybrid engine works by using both gasoline and an electric motor.

CUTTING EDGE: TURNING DOWN THE HEAT

Many projects are planned by scientists around the world to help reduce climate change.

1. Scientists in Massachusetts and Pennsylvania are building special water-treatment plants that enable oceans to absorb more carbon dioxide from the atmosphere.
2. British scientists James Lovelock and Chris Rapley hope that dotting the world's oceans with 200-m (655-ft.) tubes to bring nutrients from the deep up to the surface will encourage algae to grow. The idea is that the algae would suck carbon dioxide out of the atmosphere. If their calculations are correct, they plan to install networks of pipes in the Gulf of Mexico and the Coral Sea off northeastern Australia.
3. American scientist John Latham, from the National Center for Atmospheric Research in Colorado, has come up with an idea to make clouds bounce more sunlight back into space and cool down Earth. He hopes to inject clouds with extra water vapor through 20-meter-high (65-feet-high) rotors aboard ships. He believes that clouds covering 3 percent of Earth's surface would balance out the warming from carbon dioxide.
4. American scientist Lowell Wood, from the University of California, is looking into firing particles into Earth's upper atmosphere. The idea is that these will reflect sunlight back into space. He has also suggested using rockets and solar sails, placed high above Earth, to screen out some of the Sun's rays.

These ideas are not as important as reducing fossil fuel use and limiting deforestation in tackling climate change. However, each project may eventually help to cool the planet in a small way.

Vertical pipes placed in the sea would move up and down with the swell of the water, helping to redistribute cold-water nutrients to the surface of the ocean. Algae and other ocean life would be more likely to grow in the cooler water and help to absorb carbon dioxide from the atmosphere.

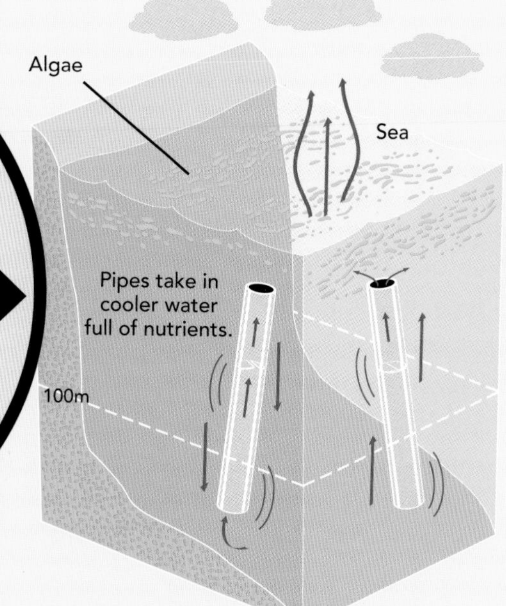

The nutrient rich water encourages growth of algae, which absorbs carbon dioxide (CO_2) from the atmosphere.

Pipes release cool water to the surface.

Individual action

If everyone on the planet made a few small changes in their daily lives, we could have a massive impact in the fight against climate change. We can never completely control Earth's natural cycles of warming and cooling, but we can try to put right the damage that humans have already caused. If everyone in the world cut down their transportation use, made small changes to their home energy use, and paid attention to the foods they eat, we could meet the targets scientists believe are needed to reduce greenhouse gas emissions.

Energy-efficient light bulbs use up to 80 percent less energy than standard bulbs. Each one could save several pounds of CO_2 in a year.

Ways to make a difference

1. Travel light. The majority of transportation still runs on fossil fuels, and each mile traveled releases carbon into the atmosphere. Would you choose an environmentally-friendly car? A large, four-wheel-drive vehicle produces about three times as much carbon dioxide as a small, four-cylinder car. Or perhaps you would consider buying a fuel-cell car? Public transportation, and trains in particular, use less fuel per person than cars. Walking and cycling are even better for our health and the planet.
2. Save energy. Electricity is the main source of power in urban areas. Our gadgets run on electricity generated mainly from power stations that burn fossil fuels.
3. Eat wisely. Growing and transporting food can be environmentally unfriendly.
   - Support local farmers. The transportation used to carry food across the world releases a lot of carbon dioxide. By curbing our appetites for exotic foods and eating local produce, we could help make the air cleaner.
   - Grow vegetables to reduce transportation and packaging, and start a compost heap to recycle nutrients into the soil, instead of using fertilizers.
   - Buy food that doesn't use much packaging.
4. Reduce waste in general, and recycle as much as possible. The fewer new products you use, the less energy is needed to make them.
5. Educate yourself and others. The more you know, the bigger the difference you can make.

IN YOUR HOME: WHAT CAN FAMILIES DO AT HOME?

Very small changes in the home can save a lot of energy. Insulating homes, turning down the thermostat, wearing an extra sweatshirt, and buying energy-efficient appliances can make all the difference to the planet—and the bills!

Try these ways to save electricity:

- Only use the amount of water you need.
- Only use washers, driers, and dishwashers when you have a full load.
- Don't leave the television or DVD player on standby.
- Switch lights off when you leave a room.
- Take showers instead of baths. The less water you use, the less it has to be purified, heated, and treated, all of which use energy.
- Replace ordinary light bulbs with energy-efficient light bulbs.

Producing plastic, glass, cans, paper, and clothes burns huge amounts of energy. Try the following simple steps to save energy.

- Recycle glass, newspapers, magazines, and tin cans. Use local recycling centers.
- Reuse plastic shopping bags, or bring your own reusable bags.
- Give unwanted gifts and clothes to a charity shop.

Using household waste to make compost not only enriches soil, but also helps it to store carbon, according to some scientists.

Carbon footprints

The next few decades will see many changes in the way we travel, make and use our energy, and deal with the excess carbon dioxide in the atmosphere. Science will come up with many more ideas and answers, but each of us need to reduce our own **carbon footprint**. A carbon footprint is a figure for a person's individual carbon dioxide output. It is usually measured in tons as a yearly total. Websites with carbon calculators use your information about annual car and air travel and monthly power usage to give a measure of your carbon footprint. If everyone on the planet reduced their carbon footprint, Earth would be in less danger from climate change.

Cutting the carbon

Carbon capture and storage (CCS) is a new technology that involves capturing carbon dioxide from coal or gas power plants. By storing carbon dioxide deep underground, we reduce the amount released into the atmosphere. CCS could probably reduce emissions from a power station by 70–80 percent, but there are presently no CCS coal-fired power stations operating in the world. The technology has not yet been proven to work, and it might be too expensive to operate. However, it is hoped that one day it could help to reduce our carbon footprint.

Carbon footprints refer to the amount of carbon dioxide each of us leaves behind just by going about our daily lives.

Scientists are now developing clean coal for use in power stations. This involves gasifying coal before it is burned. The gasifying process removes the carbon dioxide, which makes coal a much cleaner fuel. This technique is already being used in power stations in the United States and China.

There is still a long way to go and new technologies to discover before the problem of climate change is solved. Everyone has to play their part in tackling the problem. In the future, scientists must help to reduce the causes of climate change, but also to come up with ways for people to cope with future changes in our climate. Never has science had such a major part to play in gathering information, interpreting data, and designing solutions to save our planet. The science you learn in the classroom has everything to do with survival in the real world, and proves that science really does matter.

The average carbon footprints for different countries in 2004

Country	Position in world	Metric tons per person per year
Qatar	1st	69.2
United States	10th	20.4
Canada	11th	20.0
Australia	13th	16.3
Germany	37th	9.79
United Kingdom	37th	9.79
Netherlands	43rd	8.74
New Zealand	50th	7.8
France	63rd	6.2
China	91st	3.84
Brazil	120th	1.80
India	133rd	1.20
Zambia	179th	0.20
Chad	206th	0.01

Facts and Figures

Climate change timeline

For more than 100 years, scientists have been concerned about human activity affecting the climate.

1895 Swedish scientist Svante Arrhenius describes what comes to be known as the greenhouse effect. He suggests that doubling the amount of carbon dioxide in the atmosphere might raise Earth's temperature by between 5 and 6°C (9–11°F).

1913 American scientist Charles Abbot of the Smithsonian Astrophysical Observatory studies sunspot cycles and weather patterns, and provides evidence of the first link between solar activity and weather.

1927 Yugoslavian Milutin Milankovic proposes that small, naturally occurring changes in Earth's orbit affect climate, causing ice ages and warm periods. Records later support this theory.

1956 Canadian scientist Gilbert Plass researches climate and analyzes how carbon dioxide traps heat. He states that human activity could raise the average global temperature, and the world takes notice.

1971 Climate change on Mars is observed by the *Mariner 9* spacecraft. The red planet is enveloped by a massive dust storm. The dust absorbs solar radiation, which warms the surface by tens of degrees Celsius.

1979 The First World Climate Conference is held in Geneva, Switzerland. Deforestation and ecological changes are recognized as contributors to climate change.

1982 Ice cores from the Greenland ice sheet show dramatic temperature changes. Scientists report strong evidence for global warming and call 1981 the warmest year on record.

1988 The United Nations establish the Intergovernmental Panel on Climate Change (IPCC). Carbon dioxide concentrations in the atmosphere reach 350 ppm.

1990 The IPCC presents its First Assessment Report, which states that human activities are increasing greenhouse gases in the atmosphere.

1992 The United States, along with more than 100 other countries, sign the United Nations Framework Convention on Climate Change in Rio de Janeiro, Brazil, which aims to reduce emissions of greenhouse gases.

1997 The Kyoto Protocol is agreed upon to reduce collective greenhouse gas emissions to an average of 5.2 percent below 1990 levels by 2012. The EU agreed to a target of 8 percent below the 1990 level.

2001 President Bush (U.S. president 2001–09) withdraws the United States from the Kyoto treaty. The United States National Academy of Sciences publishes a report on the human role in climate change. The panel declares "temperatures are, in fact, rising" and states that human activity is the likely culprit. Predictions of a 3°C (5°F) warming are called "consistent" with climate science.

2005 The Kyoto Protocol comes into force, agreed upon by more than 140 countries. Concentration of carbon dioxide now stands at 372 ppm, higher than at any time in at least the last 420,000 years.

2007 The Nobel Peace Prize is awarded to international scientists "for their efforts to build up and disseminate greater knowledge about man-made climate change, and to lay the foundations for the measures that are needed to counteract such change."

2007 The Bali Roadmap Agreement sets an action plan to reduce carbon emissions across the world.

2008 Head of the United Nations climate panel states that the developed world should help poor countries prepare for global warming by assisting them in restoring coastal forests and in training health care workers for inevitable changes.

[Source: CNN]

Fast facts

- The spring ice thaw in the northern hemisphere occurs 9 days earlier than it did 150 years ago, and the autumn freeze now typically starts 10 days later.
- The 1990s was the warmest decade since the mid-1800s, when record keeping began. However, it seems likely that the 2000s will finish hotter still. The hottest years recorded are 1998, 2005, 2002, 2003, 2001, and 1997.
- By the year 2050, up to one million additional deaths from malaria may occur annually as a result of climate change.

For the record

Venus has undergone massive greenhouse warming. Its atmosphere is 96 percent carbon dioxide. This traps solar radiation and heats the planet's surface to an average temperature of 467°C (872°F). That's hot enough to melt lead!

Can you believe it?

Robert Mather, a scientist at Heriot-Watt University in Edinburgh, is developing solar-powered fabrics. The aim is to create solar cells that can bend as the fabric moves, creating flexible solar panels and possibly even solar clothes. "We envisage these new solar cells on curved surfaces of buildings, or as transportable power that can be rolled up and moved around," he says.

El Niño

Every few years, a climate pattern called El Niño takes place. The water near the equator in the Pacific Ocean gets hotter than usual and affects the atmosphere and weather around the world. El Niño climate conditions are not predictable.

El Niño can change the weather of the United States, particularly in California and the southern states. Usually, El Niño brings more rain and higher temperatures. In an El Niño year, warm ocean currents reach farther north and all kinds of tropical fish are then found in the waters far north along the west coast of the United States. El Niño can also bring warmer-than-normal winter temperatures to the eastern part of the United States.

Although it is impossible to predict the effects of global warming on the frequency of El Niño events, all indications seem to be that the effects are becoming stronger, more common, and are no longer disappearing completely in between El Niño years.

Find Out More

Books

Dow, Kristin and Thomas E. Downing. *The Atlas of Climate Change: Mapping the World's Greatest Challenge*. Berkeley, CA: University of California Press, 2006.

Gore, Al. *An Inconvenient Truth: The Planetary Emergency of Global Warming and What We Can Do About It*. Emmaus, PA: Rodale Books, 2006.

Henson, Robert. *The Rough Guide to Climate Change*. New York: Rough Guides, 2008.

Langholz, Jeffrey and Kelly Turner. *You Can Prevent Global Warming (and Save Money!): 51 Easy Ways*. Kansas City, MO: Andrews McMeel Publishing, 2008.

Websites

http://tiki.oneworld.net/global_warming/climate_home.html
This site includes a lot of facts on global warming issues.

http://www.climatehotmap.org/
A map illustrates the local consequences of global warming.

http://www.ncdc.noaa.gov/oa/climate/globalwarming.html
The National Oceanic and Atmospheric Administration National Climatic Data Center answers frequently asked questions about global warming.

http://www.ucsusa.org/global_warming/solutions/what-you-can-do-about-global-warming.html
Find out more about what you can do about global warming.

www.epa.gov/climatechange/kids
A lot of background information on climate change from the United States Environmental Protection Agency.

www.nasa.gov/vision/earth/everydaylife/climate_class.html
See how Earth's climate may change by 2060.

www.unep.org/Themes/climatechange/PDF/factsheets_English.pdf
Detailed fact sheet on climate change from the United Nations.

Film

An Inconvenient Truth—documentary DVD and further information available at www.climatecrisis.net

Watch the World Wildlife Fund Climate Change film on:
http://www.wwf.org.uk/arcticappeal/arcticvideo.asp

Topics to research

- The effect of the Gulf Stream on Europe's climate
- Electric car design and car production in the developing world
- Air transportation and global warming
- Antarctica research and surveys
- NASA climate research

Glossary

aerosol mixture of fine solid or liquid particles and gas

alga plant or plant-like organism (such as seaweed) that grows in water. Algae is the plural of alga.

atmosphere layers of gases surrounding Earth

biofuel fuel from plant and animal resources, especially plants, vegetable oils, and treated wastes

carbon element found everywhere on Earth, from diamonds, coal, and oil, to the bodies of living things

carbon footprint measure of the impact that human activities have on the environment in units of carbon dioxide

chlorophyll green pigment in plants that is necessary for photosynthesis

Clean Development Mechanism projects plan for developed countries to invest in programs to reduce greenhouse gas emissions in developing countries. By doing this the developed country can earn "credits" to offset (balance) their own greenhouse gas emissions.

ecosystem living things interacting with their environment under natural conditions

evaporation change from a liquid to a gas

food chain series of organisms showing the order in which one uses the next as a food source

fossil fuel fuel such as coal, oil, or natural gas that is formed in Earth from plant or animal remains over millions of years

geothermal heat from underground, inside Earth

global warming heating of Earth as a result of major air pollution and deforestation

greenhouse gas gas in the atmosphere that contributes to the greenhouse effect that warms Earth

hybrid something of mixed origin or composition, such as a car fueled by both gasoline and electrical energy

hypothesis unproven theory assumed to be true for purposes of argument or further study and investigation

ice age long period of widespread freezing with large areas of Earth covered by thick ice

NASA National Aeronautics and Space Administration. NASA is an independent government agency that was set up in 1958 to research and develop space exploration.

ozone layer part of Earth's atmosphere that blocks most of the Sun's harmful rays

photosynthesis process in which plants take in carbon dioxide from the air and use sunlight to make their own food. Photosynthesis produces oxygen.

phytoplankton tiny (often microscopic) plant organisms that float in ocean water

radar method of sending out radio waves and measuring their reflection to gain information

renewable energy energy source that can be replaced by natural processes, such as solar, wind, and water energy

sediment tiny pieces of soil, animal remains, and rock

sedimentary formed by or from sediment

simulation imitation by a system or process of the way in which another system or process works

smog thick haze caused by the action of sunlight on air polluted by smoke and fumes

solar energy energy from the Sun

transpiration process by which moisture is released from plants and trees

ultraviolet radiation invisible rays that are part of the energy that comes from the Sun

United Nations international organization with over 190 member countries, all agreeing to cooperate in international law, security, economic development, and human rights

variable detail, amount, or information that is likely to change

Index